Arches, Arches Everywhere!

by Allison K. Lim

Look at the land. What do you see? Arches, arches everywhere! This is Arches National Park. People didn't carve these rocks. Nature did!

Arches National Park

Utah

N
W   E
S

Long ago, there were no arches here.
There were large chunks of rock.
Wind and water caused weathering
and erosion.

Weathering and erosion changed the rock. Over time, the rocks changed shape. They turned into arches.

Double O

This is Double O Arch. Wind and rain carved two circles in this rock. The big arch is easy to see. But a much smaller arch is below it.

These are The Spectacles. Spectacles are glasses. Wind and rain carved two openings in this rock. The openings are shaped like eyes.

This is Landscape Arch. It is the largest arch in the park. Wind and rain made this arch long and skinny.

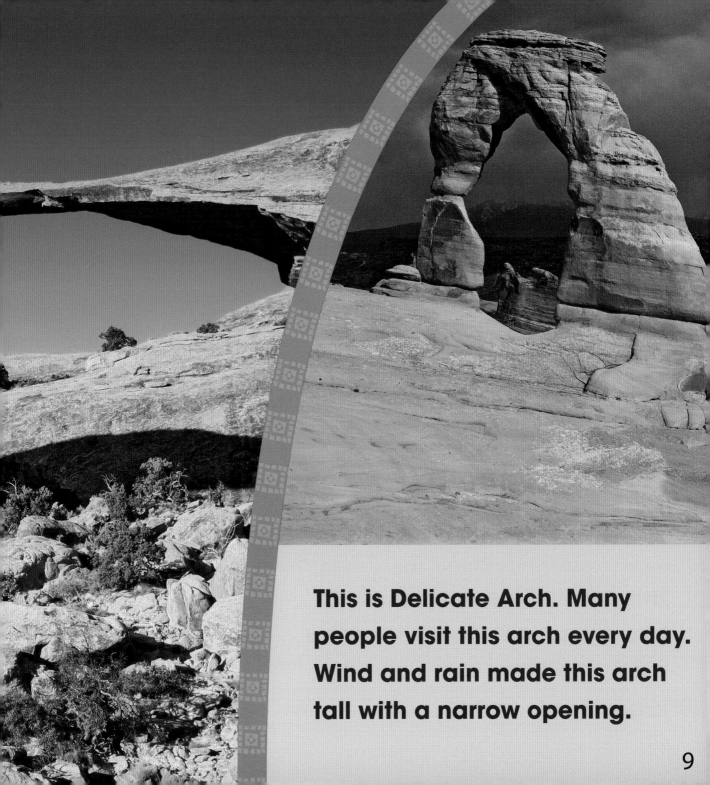

This is Delicate Arch. Many people visit this arch every day. Wind and rain made this arch tall with a narrow opening.

This is Pine Tree Arch. A pine tree grows in the middle of this arch. Wind and rain carved a jagged hole in this rock.

This is **Sand Dune Arch**. It sits in a pile of deep, reddish sand. Wind and rain carved a smooth hole in this rock.

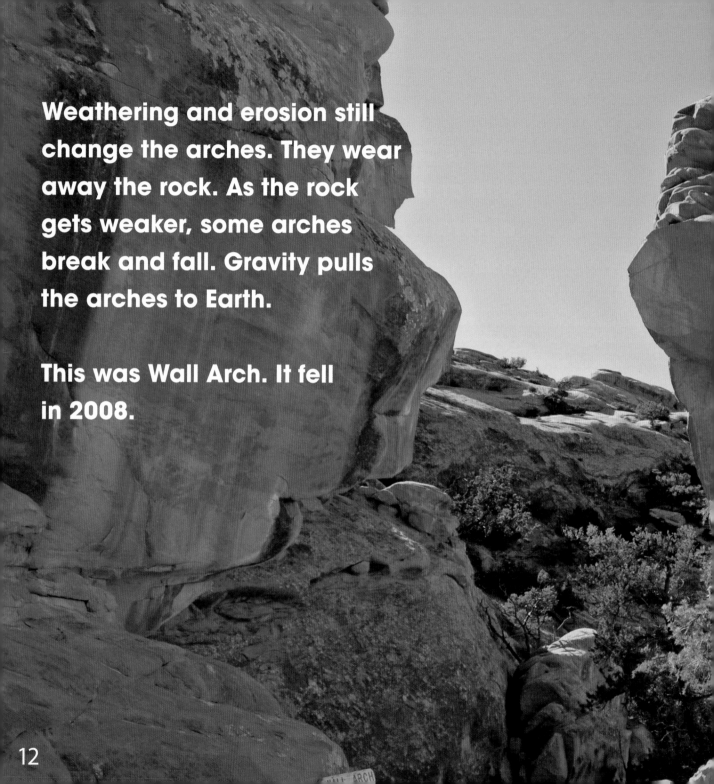

Weathering and erosion still change the arches. They wear away the rock. As the rock gets weaker, some arches break and fall. Gravity pulls the arches to Earth.

This was Wall Arch. It fell in 2008.